19 GALAXIES CAPTURED

BY

JAMES WEBB TELESCOPE

JWST Depicts Staggering Structure in Nineteen Celestial Bodies

JOHN J. MARTINEZ

Copyright ©2024 JohnMartinez

TABLE OF CONTENTS

Credit: NASA, ESA, CSA, STScI, Janice Lee (STScI), Thomas Williams (Oxford), PHANGS Team

Each of the 19 images is displayed in high resolution and thoroughly described in Chapter 2

INTRODUCTION

19 spiral galaxies have stunning new views made possible by the James Webb Space Telescope, which has also provided previously unheard-of insight on their secrets. With the help of Webb's remarkable capacity to view in various infrared light wavelengths, these stunning photos have allowed us to get up close and personal with the millions of stars, whirling gasses, and complex dust formations that characterize these celestial giants.

As part of the PHANGS (Physics at High Angular resolution in Nearby GalaxieS) study, Webb made some astounding findings that have stunned scientists. Upon facing each galaxy, one can see spiral arms studded with brilliant stars, and their cores contain mysterious supermassive black holes and groupings of old stars. However, that's just the start.

The most important part of this revelation is presented in Chapter Two, when all 19 pictures are fully shown and explained. One by one, you will travel through these galaxies and see the details of their spiral arms, star clusters in their centers, and the colorful gas and dust that envelops them.

We can see millions of dazzling blue stars grouped together and dispersed across the spiral arms of these galaxies when we look into their centers with the help of Webb's Near-Infrared Camera. The Mid-Infrared Instrument onboard Webb shines light on the bright dust that surrounds these stars, revealing red, young stars encased in the same substance that fuels their development.

The orange and red gas in the spiral arms of these galaxies is blazing, giving important information about how gas and dust are distributed throughout these celestial marvels. Through his photos, Webb sheds light on how galaxies simultaneously promote and inhibit star formation, providing an intimate look into the complex dance of star creation.

However, the wonders don't stop there. Massive, spherical voids created by the violent demise of stars have also been caught by Webb's lens among the galactic gas and dust. These mysterious emptiness bear witness to the ever shifting character of the cosmos.

We can piece together the history of these galaxies' development by studying their anatomical features, which begin at the galactic center and spiral outward. The galactic population's age is indicated by the number of blue stars close to the center of the galaxy, while younger stars are found farther from it.

Awe-inspiring and breathtakingly intricate, the pictures captured by Webb's lens are explained in detail in Chapter Two, and they will not only deepen our comprehension of these galaxies but also leave you in awe. We shall delve further into the mysteries and tales these pictures depict about the starry worlds beyond our planet's boundaries as we examine them in great depth in the pages that follow.

CHAPTER ONE

VIEWING VIA WEBB'S LENS

Synopsis of JWST's Infrared Astronomy

The First Views of a New Audience

The James Webb Space Telescope (JWST), perhaps the most sophisticated observatory to date, was launched by humanity in an attempt to peer further into space. JWST represents the curiosity and desire of a human race ever searching for greater heights. It was designed to be the first post-Hubble observatory. The telescope was the result of several years of meticulous engineering, planning, and international cooperation between NASA, the European Space Agency (ESA), and the Canadian Space Agency (CSA).

When the Hubble Space Telescope was launched in the early 1990s, JWST was initially suggested. The JWST was designed to be the trailblazer of infrared astronomy, with the understanding that there was much more to be found in wavelengths beyond visible light. After overcoming both financial and technical obstacles, JWST was finally launched on December 25, 2021, from French Guiana using an Ariane 5 rocket.

An Infrared Portal to the Cosmos

JWST unfurled its massive sunshield and gold-coated mirrors upon reaching the second Lagrange point (L2), and embarked on its mission to unravel the mysteries of the cosmos using infrared light. Still, why infrared? The answer lies in the universe's own expansion. As the cosmos expands, redshift is the process by which light from the first stars and galaxies shifts from the visible to the infrared region of the spectrum. Furthermore, infrared light might act as a cosmic clarifier by passing through clouds of cosmic dust that obscure a multitude of celestial bodies.

The Infrared's Boundary

The instruments in JWST's suite, including the Mid-Infrared Instrument (MIRI), the Near-Infrared Spectrograph (NIRSpec), and the Near-Infrared Camera (NIRCam), provide an unmatched perspective into the universe's hidden sides. By collecting images at various infrared wavelengths, these gadgets provide a complete view of the universe.

The first galaxies are seen as NIRCam sharpens the vision of the distant universe. NIRSpec analyzes the composition of cosmic objects by dispersing an object's light throughout a spectrum. Through the mid-infrared band, MIRI expands JWST's reach, enabling it to study colder objects in space, such planetary systems forming around young stars and the dust-covered cradles of star formation.

JWST's strength comes from its ability to collect images at many infrared wavelengths. Different wavelength bands display different aspects of the cosmos. Shorter wavelengths bring out the details of the stars and galaxies in the images, making them crisper. Larger gas and dust clouds that are giving birth to new stars are examples of objects with lower temperatures that are highlighted by longer wavelengths. JWST's multi-layered view of the universe, made possible by these capabilities, enables investigation of the cosmic landscape that was before impossible.

Not only is JWST an observatory, but it also functions as a time machine, capturing the faint whispers of light that have been moving across space for billions of years. Through this lens, we may see the emergence of the earliest galaxies, stars forming in murky nurseries, and the blueprints of planetary systems that could harbor life itself. We are getting closer to finding the answers to long-standing questions regarding the origins, evolution, and fundamental components of the universe with each image and spectrum that JWST takes on its ongoing infrared trip.

The PHANGS Collaboration

Bring the Community of Astronomers Together! Understanding the genesis and development of galaxies in the vastness of space is a complex task. The Physics at High Angular resolution in Nearby Galaxies (PHANGS) project is an ambitious endeavor to bring these pieces together. Through this alliance, over a hundred astronomers from across the globe are collaborating to solve the riddles of star formation and galaxy history, with everyone contributing their own area of expertise.

The Objective of PHANGS

PHANGS aims to map nearby galaxies with unprecedented depth of information on their structure and composition. PHANGS focuses on galaxies that are somewhat close to our own Milky Way in an effort to provide light on the processes driving star creation and the dynamics that shape spiral galaxies.

Power of High-Resolution Photos

Resolving complicated characteristics seen inside galaxies requires high-resolution measurements. With the use of the most advanced observational technology, PHANGS integrates data from several sources. Dust-covered star formation regions are illuminated by the James Webb Space Telescope (JWST), which contributes significantly with its unparalleled infrared imaging capabilities. The Atacama Large Millimeter/submillimeter Array (ALMA) offers complementary views in the radio spectrum by tracking the cold molecular gas that forms the nucleus of young stars. Meanwhile, the European Southern Observatory's Very Large Telescope (VLT) and Multi Unit Spectroscopic Explorer (MUSE) instrument give spectral data that maps out the star populations and ionized plasma.

The way these telescopes cooperate, each seeing a different piece of the cosmic puzzle, is what makes the PHANGS project so groundbreaking. The benefit of collaboration in scientific activities is shown by the fact that data from several wavelengths are integrated to provide a full picture. Using a mix of simulations, theories, and observations, the multidisciplinary team behind the project closely investigates this data to challenge our understanding of stellar physics.

The Vital Role of Collaboration

The significance of the PHANGS cooperation is multifaceted. It first does this by using the unique qualities of many observatories to create a multi-wavelength approach that provides a more complete image of galaxies. Its ability to convene a diverse range of expertise, such as data analysts, observers, and theorists, provides a platform for the testing and debate of new theories, which is its second advantage. Not to mention, PHANGS is setting a precedent for future large-scale astronomy project execution and highlighting the many advantages of international cooperation.

Combining the infrared vision of JWST with the capabilities of ALMA and VLT, PHANGS is rewriting the tale of how stars and planetary systems emerge in addition to studying galaxies. As a guide for collaborative scientific inquiry, this alliance points the way toward discoveries that would be unattainable for any one observatory or research team acting alone.

Picture Capture: An Amazing Technological Device

Making Use of the Infrared Spectrum

To reveal the hidden beauty of the cosmos, the James Webb Space Telescope's array of cutting-edge instruments, each focused on a specific area of the infrared spectrum, is essential. Two such jewels in Webb's crown that provide astronomers an unmatched view into space are NIRCam (Near-Infrared Camera) and MIRI (Mid-Infrared Instrument).

NIRCam: The Explorer of the Stars

NIRCam, Webb's primary imaging instrument, was designed to examine the universe in the near-infrared, or between 0.6 and 5 micrometer, wavelength range. NIRCam, designed to look through cosmic dust and capture the light of the earliest stars and galaxies, is a crucial tool for addressing some of the most significant issues in contemporary astronomy. With its state-of-the-art optics and detectors, it is able to detect even the tiniest light particles that have traveled across space for almost 13 billion years.

Uncovering the Universe's Dust Cycle with MIRI

Beyond 5 to 28 micrometers in the infrared, MIRI extends Webb's field of view into the colder, dustier regions of space where stars and planetary systems form. Unlike NIRCam, MIRI is able to examine the chemical fingerprints of cosmic objects in addition to obtaining detailed images since it is equipped with a spectrograph. Because of MIRI's ability to recognize and identify organic molecules and other substances, it is conceivable to comprehend the composition of planets and the potential for life elsewhere in the universe.

The Challenges and Triumphs of Imaging

The JWST's imaging technology faced several challenges never seen before in its development and deployment. The apparatus had to be constructed to withstand the harsh circumstances of space travel and operate in the severe conditions of the L2 point, when temperatures are almost zero. The primary mirror's eighteen hexagonal components required precise engineering because they had to line properly after unfolding in space—a procedure that had never been attempted before.

The mission was deemed successful when JWST released its "first light" photos. These first views of space were so clear and profound that they stunned the scientific community as well as the general people. These images served as a tribute to the telescope's capabilities and a big step forward for space exploration.

Being able to detect infrared light from space with NIRCam and MIRI has been a significant achievement in scientific equipment. The technology of these sensors has allowed us to glimpse the early stages of the universe, something that was previously unimaginable. They provide views of everything from the weather on exoplanets to the formation of distant galaxies that have never been seen before.

The first pictures that Webb and its equipment took were more than just pictures; they were the product of decades of imagination and perseverance. With the help of the most advanced observatory ever constructed, every image is brought to life like a tapestry created from countless photons that have traveled across time and space. Scientists are still enthused by these images, and we have only just started to investigate the potential that each one has to change our understanding of the universe.

CHAPTER 2

A HARMONY OF STARS

Thorough Descriptions of Each of the 19 Spiral Galaxies, Emphasizing the Amazing Images That Webb Captured

1. NGC 5332

Image Credit: NASA, ESA, CSA, STScI, Janice Lee (STScI), Thomas Williams (Oxford), Rupali Chandar (UToledo), PHANGS Team

Hubble is at bottom right and Webb is at top left in this diagonal arrangement of the two images of the galaxy IC 5332. The galactic core is roughly centered, yet the galaxy's arms seem to spin in a clockwise direction. The spiral arms in Webb's image have a spine-like appearance and are composed of many orange-colored filaments with distinct "bubbles" of dark gray or black and a blue haze in the middle. While pink star-forming areas, light brown dust lanes, and bright blue star clusters make up the spiral arms, the picture's center is a soft yellow color captured by Hubble's camera.

2. NGC 628

Image Credit: NASA, ESA, CSA, STScI, Janice Lee (STScI), Thomas Williams (Oxford), Rupali Chandar (UToledo), PHANGS Team

Two diagonally split observations of a portion of NGC 628 are shown, with Webb's at top left and Hubble's at bottom right. The galactic core is located roughly in the center, whereas the galaxy's arms seem to rotate counterclockwise. Spiraling filament structure is similar to a nautilus shell cross section in several aspects. In Webb's image, the spiny spiral arms consist of several orange-colored filaments with observable dark gray or black "bubbles," and a blue haze surrounds the center. Hubble's image has a pale yellow core, with pink star-forming areas, dark brown dust lanes, and brilliant blue star clusters arranged in a spiral pattern around it.

3. NGC 1087

Image Credit: NASA, ESA, CSA, STScI, Janice Lee (STScI), Thomas Williams (Oxford), Rupali Chandar (UToledo), PHANGS Team

The photos of the galaxy NGC 1087 taken by Webb and Hubble are positioned at the top and bottom of a diagonal plane, respectively. The galaxy's core is located roughly in the center, and its arms rotate in a clockwise direction. It might be difficult to distinguish between the spiral arms since they seem to be blended together. The core has a slender line-like form and has a dazzling yellow color, as seen in Webb's picture. The spiral arms consist of several orange filaments that have pronounced dark gray or black "bubbles." The spiral arms in Hubble's image are composed of bright blue star clusters, pink star-forming areas, and dark brown dust lanes, with a soft yellow hue occupying the middle.

4. NGC 1300

Image Credit: NASA, ESA, CSA, STScI, Janice Lee (STScI), Thomas Williams (Oxford), Rupali Chandar (UToledo), PHANGS Team

Two diagonally split pictures of the NGC 1300 galaxy are shown, with Webb's image at top left and Hubble's at bottom right. The galaxy's center is connected to a prominent diagonal bar structure. The two distinct spiral arms of the galaxy begin to turn counterclockwise at the end of the bar. A reverse S shape is formed when the arm and bars are combined. The round, light yellow core portion in Webb's image supports the bar and spiral arms, which are composed of many orange filaments in a variety of hues. Hubble's image's bar and center are a gentle yellow, while the spiral arms are made up of a sparkling blue star cluster and a combination of dark brown dust lanes.

5. NGC 1365

Image Credit: NASA, ESA, CSA, STScI, Janice Lee (STScI), Thomas Williams (Oxford), Rupali Chandar (UToledo), PHANGS Team

The two observations of a portion of NGC 1365 are placed diagonally, with Hubble's observation at top left and Webb's at bottom right. The bar structure of the galaxy extends horizontally, like a reverse S with a little curvature, and its core can be seen to the left, somewhat above center. In Hubble's image, the bar and the beginning of the spiral arms are made up of a combination of dark brown dust lanes and a few bright blue star clusters, while the galaxy's core is centered around a luminous yellow dot. Orange tones, distinct dark gray or black "bubbles," and a blue haze around the bar and core are all present in the dust lanes that cross the bar in Webb's image.

6. NGC 1385

Image Credit: NASA, ESA, CSA, STScI, Janice Lee (STScI), Thomas Williams (Oxford), Rupali Chandar (UToledo), PHANGS Team

Two diagonally split views are shown of the galaxy NGC 1385, with Webb's at top left and Hubble's at bottom right. The galaxy has a roughly defined core. It is difficult to identify individual spiral arms from the chaotic look of the galaxy. In Webb's image, the spiral arms are jumbled, with pronounced "bubbles" of dark gray or black amid the orange tones. In Hubble's image, the tangled spiral arms are composed of pale yellow, dark brown dust lanes, and bright blue star clusters.

7. NGC 1433

Image Credit: NASA, ESA, CSA, STScI, Janice Lee (STScI), Thomas Williams (Oxford), Rupali Chandar (UToledo), PHANGS Team

The photos of the galaxy NGC 1433 taken by Webb and Hubble are positioned at the top and bottom of a diagonal plane, respectively. A large diagonal bar structure, associated with the oval center of the galaxy, occupies most of the frame. Starting at the bar's end, the galaxy's arms seem to spin counterclockwise. The spiral arms in Webb's image are composed of many filaments in different shades of orange. Thin dust lanes emerge from the core via the spiral arms and the bar. In Hubble's perspective, the delicate blue heart of the spiral arms contrasts with dark brown dust lanes and blue star clusters on the outskirts.

8. NGC 1512

Image Credit: NASA, ESA, CSA, STScI, Janice Lee (STScI), Thomas Williams (Oxford), Rupali Chandar (UToledo), PHANGS Team

The two perspectives of NGC 1512 are split diagonally, with Webb's view at top left and Hubble's at bottom right. The center of the galaxy is round, and in the upper left and lower right corners, two prominent dust lanes are seen as they traverse the huge diagonal bar structure. The top blue bar seems to be a gap at first, but closer examination shows that it feeds a larger circular shape that has several spiral arms, like a braid. In Webb's image, the spiral arms are composed of many filaments with different shades of orange. In Hubble's perspective, the spiral arms are composed of bright blue star clusters and dark brown dust lanes, with a faint pinkish-brown bar.

9. NGC 1566

Image Credit: NASA, ESA, CSA, STScI, Janice Lee (STScI), Thomas Williams (Oxford), Rupali Chandar (UToledo), PHANGS Team

The two perspectives of NGC 1566 are placed diagonally, with Hubble's view at top left and Webb's at bottom right. The galaxy's nucleus seems to be centered, while its arms rotate counterclockwise. Light yellow makes up the center of Hubble's image, while two prominent spiral arms are made up of dark brown dust lanes and a stunning blend of pink and blue star clusters. In Webb's shot, the spiral arms have orange tones with pronounced "bubbles" of dark gray or black and a blue haze around the center.

10. NGC 1672

Image Credit: NASA, ESA, CSA, STScI, Janice Lee (STScI), Thomas Williams (Oxford), Rupali Chandar (UToledo), PHANGS Team

The two perspectives of NGC 1672 are placed diagonally, with Hubble's view at top left and Webb's at bottom right. Only in the Webb region of the image, to the right of center, is the galaxy's core visible. A almost horizontal bar structure connects two prominent spiral arms that seem to be rotating clockwise. Hubble's image of the spiral arms consists of a combination of bright blue star clusters and dark brown dust lanes. In Webb's image, the bar and spiral arms are made up of several orange filaments that have observable dark gray or black "bubbles." The core of the galaxy is a big yellow circle with a blue dot in the middle.

11. NGC 2835

Image Credit: NASA, ESA, CSA, STScI, Janice Lee (STScI), Thomas Williams (Oxford), Rupali Chandar (UToledo), PHANGS Team

Two observations of NGC 2835 are shown, one at the top left by Webb and the other at the bottom right by Hubble. The galaxy's nucleus is centered, yet its arms seem to rotate counterclockwise. Despite their chaotic appearance, the spiral arms may be identified individually. In Webb's image, the center is bright blue, while the spiral arms are composed of many orange-colored filaments with distinct "bubbles" of dark gray or black. Hubble's image shows a soft yellow core surrounded by spiral arms made up of a combination of dazzling blue star clusters and dark brown dust lanes.

12. NGC 3351

Image Credit: NASA, ESA, CSA, STScI, Janice Lee (STScI), Thomas Williams (Oxford), Rupali Chandar (UToledo), PHANGS Team

The two perspectives of NGC 3351 are placed diagonally, with Hubble's view at top left and Webb's at bottom right. The galaxy's core is located in its center, while its arms rotate clockwise to form a distinct outer ring. About half of the massive circular ring of the galaxy is made up of a combination of dark brown dust lanes and bright blue star clusters that make up the spiral arms. Hubble's image of the galaxy shows a pale yellow coloration of its almost horizontal bar structure. The other half of the large ring, which is made up of many orange-colored filaments, is represented by the spiral arms in Webb's image. Their anchor is its middle portion, a hazy blue bar that links to the core. A bigger, rounder light yellow circle with a blue dot in the center is surrounded by a smaller orange oval.

13. NGC 3627

Image Credit: NASA, ESA, CSA, STScI, Janice Lee (STScI), Thomas Williams (Oxford), Rupali Chandar (UToledo), PHANGS Team

The two perspectives of NGC 3627 are placed diagonally, with Hubble's view at top left and Webb's at bottom right. The core of the galaxy, which is shown slightly above center, is where most of Hubble's observations are found. In Hubble's image, the galaxy's core and diagonal bar structure are seen as a distinct oval that is light yellow in hue. The spiral arm that arcs across the top portion of the image is composed of a combination of dark brown filamentary dust lanes and vivid blue star clusters. In Webb's image, the spiral arm is colored in several shades of orange and stretches all the way to the bottom.

14. NGC 4254

Image Credit: NASA, ESA, CSA, STScI, Janice Lee (STScI), Thomas Williams (Oxford), Rupali Chandar (UToledo), PHANGS Team

Hubble's view of NGC 4254 is at bottom right, while Webb's view is at top left, oriented diagonally. The galaxy's nucleus is situated to the right of center, and its arms seem to spin counterclockwise. In Webb's image, the spiral arms consist of many orange-colored filaments with observable "bubbles" of dark gray or black, while the area around the core is hazy blue. In Hubble's image, the core is a mellow yellow with brown threads woven throughout, while the spiral arms are a blend of dazzling blue and dark brown. In the bottom right corner, a black triangle denotes the lack of data.

15. NGC 4303

Image Credit: NASA, ESA, CSA, STScI, Janice Lee (STScI), Thomas Williams (Oxford), Rupali Chandar (UToledo), PHANGS Team

Two diagonally split images of a portion of the galaxy NGC 4303, one taken by Webb (top left) and the other by Hubble (bottom right). The galaxy seems to have a center of gravity at its far-right edge, with sliced arms that rotate counterclockwise. A dense cluster of orange-colored filaments with distinct "bubbles" of dark gray or black make up the spiral arms in Webb's image. The spiral arms of Hubble's picture show bright blue star clusters, pink star-forming regions, dark brown dust lanes, and a light yellow center.

16. NGC 4321

Image Credit: NASA, ESA, CSA, STScI, Janice Lee (STScI), Thomas Williams (Oxford), Rupali Chandar (UToledo), PHANGS Team

Two diagonally split views of a portion of the galaxy NGC 4321 are shown, with Webb's image at bottom left and Hubble's at top right. The galaxy seems to have a center of gravity at its far-right edge, with sliced arms that rotate counterclockwise. In Webb's image, the spiral arms consist of many orange-colored filaments that have distinct black or dark gray patches. Bright blue and dazzling blue make up the picture's center, while bright blue, pink, and dark brown dust lanes make up the spiral arms.

17. NGC 4535

Image Credit: NASA, ESA, CSA, STScI, Janice Lee (STScI), Thomas Williams (Oxford), Rupali Chandar (UToledo), PHANGS Team

Hubble's image is at top right and Webb's is at bottom left of this diagonal split of the two pictures of the galaxy NGC 4535. An virtually vertical bar structure connects the galaxy's two major spiral arms, which seem to travel in a clockwise orientation, to the galaxy's core. The arms and bar create an expanded S shape. In Webb's image, the spiral arms consist of several filaments in different shades of orange, with distinct patches of dark gray or black in the middle. In Hubble's perspective, the spiral arms consist of a combination of bright blue star clusters and dark brown dust lanes, with the little core of the spiral arms appearing brilliant white.

18. NGC 5068

Image Credit: NASA, ESA, CSA, STScI, Janice Lee (STScI), Thomas Williams (Oxford), Rupali Chandar (UToledo), PHANGS Team

Hubble's and Webb's images of a different region of the galaxy, NGC 5068, are divided diagonally at the top left and bottom right, respectively. Top left is where the galaxy's center is seen. The galaxy's bar is seen in Hubble's picture as a brilliant, white area at center-left. Numerous intense blue light pinpoints, bright pink patches, and wisps of dark red are dispersed across the landscape, along with less noticeable dark brown dust lanes. The galaxy's arms are shown in Webb's picture as being interspersed with dark areas and appearing in orange and red hues.

19. NGC 7496

Image Credit: NASA, ESA, CSA, STScI, Janice Lee (STScI), Thomas Williams (Oxford), Rupali Chandar (UToledo), PHANGS Team

The two perspectives of NGC 7496 are placed diagonally, with Hubble's view at top left and Webb's at bottom right. The galaxy's core is located in its center and is connected to its two spiral arms by a diagonal bar structure. It seems that they rotate counterclockwise. In Hubble's perspective, the spiral arms consist of bright blue star clusters and dark brown dust lanes, with a pale yellow dot positioned in the center of the galaxy's bar and core. Webb's shot shows orange and red spiral arms with two prominent red diffraction spikes pointing downhill. The edges of the scene are dark.

These spiral galaxies have the ability to fascinate people easily. Trace their well-defined, star-studded arms toward their cores, which might contain ancient star clusters and sometimes active supermassive black holes. A collection of these very detailed photographs of neighboring galaxies in a mix of near- and mid-infrared light were made public today by NASA's James Webb Space Telescope, which is the only instrument capable of producing such images.

The Webb photos are a component of the Physics at High Angular resolution in Nearby GalaxieS (PHANGS) program, an extensive and well-established initiative that has the backing of over 150 astronomers globally. PHANGS was already bursting at the seams with information from the Hubble Space Telescope of NASA, the Multi-Unit Spectroscopic Explorer of the European Southern Observatory's Very Large Telescope, and the Atacama Large Millimeter/submillimeter Array, including observations in ultraviolet, visible, and radio light, before Webb took these pictures. Webb's contributions in the near- and mid-infrared have added a few more pieces to the picture.

According to Janice Lee, a project scientist for strategic initiatives at the Space Telescope Science Institute in Baltimore, "Webb's new images are extraordinary." Even for scientists who have spent decades studying these identical galaxies, they are astounding. The tiniest scales ever seen in observations of bubbles and filaments provide information on the star formation cycle.

As the Webb photos began to roll in, the team's excitement quickly increased. "It seems like the level of detail in these photos always overwhelms our team—in a good way," said Thomas Williams, a postdoctoral researcher at the University of Oxford in the United Kingdom.

Observe the Spiral Arms

These blue-toned photos were taken by Webb's Near-Infrared Camera, which counted millions of stars. While some stars are dispersed along the spiral arms, star clusters are made up of groups of closely spaced stars.

The MIRI (Mid-Infrared Instrument) data from the telescope shows us where luminous dust is found around and between stars. It also highlights partially created stars, which look like brilliant red seeds at the tops of dusty peaks, still enveloped in the gas and dust that fuels their development. Prof. Erik Rosolowsky of the University of Alberta in Edmonton, Canada, said, "These are where we can find the newest, most massive stars in the galaxies."

Another item that astounded astronomers? Large, spherical shells may be seen in the gas and dust in Webb's photos. Astronomer Adam Leroy of Ohio State University in Columbus said, "These holes may have been created by one or more stars that exploded, carving out giant holes in the interstellar material."

Now follow the spiral arms to locate prolonged reddish-orange patches of gas. According to Rosolowsky, "in certain parts of the galaxies, these structures tend to follow the same pattern." "We can think of these as waves, and the way they are spaced apart reveals a lot about the distribution of gas and dust inside a galaxy." Investigating these formations will provide important light on how galaxies initiate, sustain, and end star formation.

Explore the Interior

There is evidence that galaxies expand from the inside out; star creation starts at the centers of galaxies and spirals outward along their arms. A star is more likely to be younger the farthest it is from the galactic center. On the other hand, populations of older stars are located close to the cores and seem to be lighted by a blue spotlight. What about pink-and-red diffraction spike-rich galaxy cores? A staff scientist at the Max Planck Institute for Astronomy in Heidelberg, Germany, Eva Schinnerer, stated, "That's a clear sign that there may be an active supermassive black hole." "Alternatively, the image's central region has been saturated by the brightness of the star clusters." The extraordinary number of stars that Webb resolved is a terrific place to start, but there are numerous more study directions that scientists might take with the combined PHANGS data. Leroy said, "Stars can live for billions or trillions of years." "By accurately cataloging every kind of star, we can create a more trustworthy, comprehensive understanding of their life cycles."

The PHANGS team has provided the biggest library of almost 100,000 star clusters to date in addition to these photographs right away. "Our team is not equipped to handle the vast amount of analysis that can be done with these images," Rosolowsky said. "We are eager to assist the community in order to enable all researchers to participate."

CHAPTER 3

COOPERATION IN THE FACE OF DIVERSITY

One of the universe's most amazing creative masterpieces, spiral galaxies are a vast canvas of diversity. Thanks to the James Webb Space Telescope (JWST), we can explore the intricate structure of these cosmic objects up close. All galaxies have a similar spiral structure, yet each one tells a unique story because of its size, shape, color, and secrets tucked away within its spinning arms.

The Variety in Shape and Dimension

Spiral galaxies are small, faint, or enormous, brilliant, with diameters ranging from tens to hundreds of thousands of light-years. Some, like the Whirlpool Galaxy (M51), have well defined arms that closely encircle their bulges, giving them the appearance of vast design spirals. Some seem more flocculent, with arms that are disorganized and tangled, such as NGC 2841. Subsequently, there are barred spirals, like NGC 1300, whereby the spiral nuclei are spanked by star-filled core bars that drive gas towards the center and initiate star formation.

The galaxy's history of mergers and interactions, the distribution of star populations inside it, and the dynamics of its dark matter halo are just a few of the factors that affect this variety in form. A peek into how these many structural components impact a galaxy's overall appearance and behavior may be obtained via JWST data.

An Energetic Kaleidoscope

The JWST was able to take very detailed photos of spiral galaxies, and these images may provide details about the age and makeup of the star population. The spiral arms of galaxies such as Andromeda (M31) seem blue and sapphire due to the bright blue light emitted by young, burning stars. In contrast, M104, often called the Sombrero Galaxy, has bulges made up of older, cooler stars that have a warm, reddish-gold hue.

The JWST's infrared capabilities, which bring to light the glow of cosmic dust lanes that weave across the spiral arms, let us to view beyond visible light. These dust-rich regions, which are not visible to the unassisted eye, are active sites of star formation that glow in the infrared as they give birth to new star generations.

Handling Mysteries from Space

Every spiral galaxy is different, with intricate characteristics ranging from the supermassive black holes hiding at their cores to the intricate star formation processes. Supermassive black holes are objects of continuing scientific study. Their masses range from tens of millions to billions of times that of the Sun. Thanks to observations from JWST, astronomers can better understand how these large gravitational objects affect the galaxies they exist in by employing radiation and powerful jets to influence star formation.

Star production efficiency and rate in spiral galaxies are also intensively explored. The mass of the galaxy, the density and temperature of its interstellar medium, and the galaxy's structure all have an impact on the star formation process. Critical light on these processes is provided by JWST data and images, which demonstrate the complex interactions between gas, dust, and stars that power the cosmic cycle of birth and death.

The diversity of spiral galaxies discovered by the James Webb Space Telescope demonstrates the beauty and complexity of the cosmos. With its unique dimensions, internal dynamics, and color, each galaxy contributes to our understanding of the cosmic web that connects all structures in the universe. As we dive further into these celestial beauty, we uncover more mysteries about our cosmos and our place in this vast, star-studded expanse.

The Universe's Purpose

Spiral galaxies, with their beautiful arms around blazing centers, are more than just isolated starry islands in the cosmic sea; they are vital to our understanding of the universe in its whole. Understanding the complexities of cosmic growth, the distribution of matter in the universe, and the enigmatic nature of dark matter requires investigating these galaxies. The groundbreaking discoveries enabled by the James Webb Space Telescope (JWST) make this feasible.

Spiral Galaxies: The First People in Space

In the cosmic history timeline, spiral galaxies indicate important turning points. They identify a period of galactic history where star formation, disk formation, and active galactic nuclei coexist in a precarious balance. By examining the structure, speed, and star formation rates of these galaxies, astronomers may learn more about the processes that have shaped the universe from its most ancient beginnings to the present.

The infrared data from JWST provide a window into these processes, shedding light on how stars form in environments obscured by cosmic dust and offering insights into the physics underpinning galaxy formation. By linking the birth of the first stars and galaxies to the complex structures that we observe today, these results help put the cosmic jigsaw puzzle together.

Linking Up with the Cosmic Web

Spiral galaxies are linked into the vast structure known as the cosmic web, as opposed to being scattered haphazardly around the cosmos. This intricate network of dark matter filaments, interspersed with galaxies and galactic clusters, forms the nucleus of the universe's large-scale structural organization. Spiral galaxies, like as those seen near the cosmic web, function as beacons for the dark matter structure underneath them.

Spiral galaxy structure and orientation within the cosmic web may provide important clues regarding dark matter composition and the processes that have produced the cosmos. By mapping the positions of these galaxies and analyzing their interactions with the larger cosmic infrastructure and with one other, astronomers might get insights into the properties of dark matter and its role in the evolution of the universe.

Cracking the Code on Dark Matter

Dark matter is invisible, yet it affects spiral galaxy dynamics and creation in important ways. In contrast to Newtonian physics, the rotation curves of these galaxies remain flat at distances away from the center when only visible stuff is considered. This mismatch represents one of the strongest evidence for the existence of dark matter.

JWST's ability to map out the distribution of stars and gas and to see through dust-covered regions opens up new avenues for studying dark matter, as shown by its studies of spiral galaxies. By examining the interactions between dark matter and visible matter in these galaxies, scientists hope to enhance their models of the basic structure of the universe and the fundamental physics that control it.

Solving the mysteries of the cosmos requires an understanding of spiral galaxies. These star-forming regions resemble laboratories where one may study the processes of cosmic growth, from star formation to the interactions that build the large-scale structure of the universe. In addition to expanding our knowledge of these magnificent formations, spiral galaxies are being studied by the James Webb Space Telescope, which also sheds light on the dark matter that links the universe.

With every spiral galaxy we see and analyze, we go closer to understanding some of the most significant questions about the nature of the universe, our origins, and the ultimate fate of the cosmos.

This study of the function of spiral galaxies in the cosmic environment would make use of JWST images and data in addition to theoretical models and simulations. By establishing a link between the extensive research of individual galaxies and the vastness of the cosmos, this chapter aims to demonstrate the interconnectedness of all cosmic events and the ongoing endeavor to understand the universe as a whole.

CHAPTER 4

THE IMPACT OF GAS AND DUST

This picture shows the Orion Nebula, a massive cloud of gas and dust where new stars are created.

Credit: NASA/JPL-CALTECH/STSCI

The interstellar medium (ISM), which is the gas and dust that fills the spaces left by stars in galaxies, is a cosmic stage where a remarkable dance occurs. This beautiful and elegant dance is propelled by gravity, magnetic fields, and radiation. Thanks to the James Webb Space Telescope (JWST), which displays the complex interactions between gas and dust that drive the star-birth and star-death cycle in spiral galaxies, we are able to see this dance with unprecedented precision.

The Composition of the Interstellar Medium

The gas that makes up around 99% of the ISM is helium and hydrogen, with the remaining 1% consisting of dust, which are tiny solid particles composed of heavier elements including carbon, silicon, and iron. Until the deployment of JWST, our understanding of the ISM was restricted by optical telescopes' incapacity to see through the deepest clouds of dust. We can look through these layers and show the structures that are hidden within by using JWST's infrared capabilities.

Observing the nurseries for star formation

Our understanding of the conditions that lead to star formation in the ISM has been completely transformed by the data obtained by JWST. In the cold, dense regions of gas and dust, gravity begins to pull stuff together, forming clumps that will eventually form into new stars. These regions, which are typically obscured from vision in the visible spectrum, are referred to as molecular clouds. However, they radiate a warm light in the infrared, revealing their intricate structure to the JWST's sensitive detectors.

One of the most amazing findings enabled by JWST is the detailed imaging of spiral galaxies harboring these molecular clouds. The telescope observations have shown the massive and complex filament networks inside these clouds, where heavier gas clusters point to potential star formation regions. While shock waves from dead stars force the clouds to collapse in some locations, radiation and stellar winds carve out holes and influence cloud development in others, making the process both chaotic and beautiful at the same time.

The Role of Dust in Star Formation

Dust is a crucial element in the creation of stars. The gas cools the dense regions of molecular clouds and blocks out UV radiation when it compresses under gravity, releasing heat in the process. This cooling is crucial for star formation because it permits the gas to keep condensing into ever-denser clumps.

JWST's observations have allowed for a new understanding of the role of dust in star formation. By observing the light emitted and absorbed by dust grains, scientists may determine the temperature, density, and chemical composition of areas where stars are forming. These findings are critical for creating accurate models of how stars emerge and how galaxies change over time.

Energizing the Life Cycle of the Galaxy

The galactic lifetime is driven by processes other than star formation and death, such as the dance of gas and dust inside galaxies. The heavier elements that stars produce in their cores during their development, lives, and death contribute to the ISM. Later generations of stars are impacted by this enrichment, which leads to a galaxy that is constantly changing and developing.

JWST observations, which illustrate the connections between the large-scale spiral galaxy structure and the small-scale star formation processes, have shed light on this cycle. By tracking the flow of gas and dust along the spiral arms of galaxies, we may see the breathing of the galaxy. The galaxy shrinks when stars grow, then expands as those stars die and release their material back into the interstellar medium.

James Webb Space Telescope has allowed us to see spiral galaxies from a completely new perspective: the gas and dust dance that forms them. By revealing the hidden nurseries where stars are produced and throwing light on the processes that extend galaxies' lifespans, JWST's discoveries are helping us understand the intricate dance that is the universe. A crucial part of the cosmic story that JWST is now helping to convey with never-before-seen clarity and detail is this infrared dance of creation and destruction.

CHAPTER 5

EVOLUTION OF THE GALAXIES: FROM BIRTH TO STARDUST

Examining the Process of Galaxy Formation

Galaxies may be seen as celestial paintings in the cosmic fabric, each telling a unique story about their origins and evolution. The intricate process of galaxy creation is visually explored in the fourth chapter. The narrative begins in the cosmic cradle, where the star formation cycle begins in the spiral arms and ends in the galactic center. A symphony of star creation, this inward-outward process shapes the essential character of galaxies. The James Webb Space Telescope, with its unparalleled observations, has allowed us to probe further into the nuances of this cosmic beginning and to see the celestial dance that coordinates the evolution of galaxies across vast swaths of the universe.

The Importance of Observations for Galactic Evolution

The cosmic drama of galaxies is brought to light by Webb's lens, which also emphasizes the implications for our understanding of their evolution. The observed star distribution in relation to galactic centers is thoroughly examined in this chapter, which offers valuable new insights into the dynamical relationships between stellar populations and how they are arranged spatially in galaxies. These new discoveries both challenge and enhance existing models of galaxy history, leading astronomers to reevaluate the mechanisms governing the ebb and flow of stars within these cosmic structures. The narrative delves into the implications of these groundbreaking findings, illuminating the intricate dance between star life cycles and the gravitational pull that determines a galaxy's destiny.

Lastly, Having a Broader Impact on the Astronomical Search

As astronomy draws to a close, consider the broader ramifications of these groundbreaking discoveries for the field. The findings made possible by Webb's lens have not only transformed our understanding of how galaxies form, but they also provide new directions for astronomical study. We examine how these discoveries contribute to the ongoing endeavor to unravel the secrets of the cosmos. The implications extend beyond the limits of individual galaxies to the underlying structure of the universe, providing fresh perspectives on the origins and ends of galaxies as well as enhancing our understanding of cosmic processes. The latter scenes convey the excitement of astronomers, both seasoned and inexperienced, who are inspired to continue delving into and unraveling the mysteries of the cosmos by the groundbreaking findings made by the James Webb Space Telescope.

www.ingramcontent.com/pod-product-compliance
Lightning Source LLC
Chambersburg PA
CBHW060837290526
45792CB00006BB/1963